OnBoard ACADEMICS

Sound and Light

© 2015 OnBoard Academics, Inc
Portsmouth, NH
800-596-3175
www.onboardacademics.com
ISBN: 978-1-63096-064-3

OnBoard Academic's books are specifically designed to be used as printed workbooks or as on-screen instruction. Each page offers focused exercises and students quickly master topics with enough proficiency to move on to the next level.

OnBoard Academic's lessons are used in over 25,000 classrooms to rave reviews. Our lessons are aligned to the most recent governmental standards and are updated from time to time as standards change. Correlation documents are located on our website. Our lessons are created, edited and evaluated by educators to ensure top quality and real life success.

Interactive lessons for digital whiteboards, mobile devices, and PCs are available at www.onboardacademics.com. These interactive lessons make great additions to our books.

You can always reach us at customerservice@onboardacademics.com.

Sound

Sounds You Hear

Look at these pictures.

Can you fill in the box with the sound that might be made by each illustration?

What causes sound?

When a guitar is played and a sound is made, what happens? _____

_____

_____

When a drum is used and a sound is made, what happens?

_____

_____

_____

Sounds are made by vibrations.

The guitar is silent.

The guitar is playing.

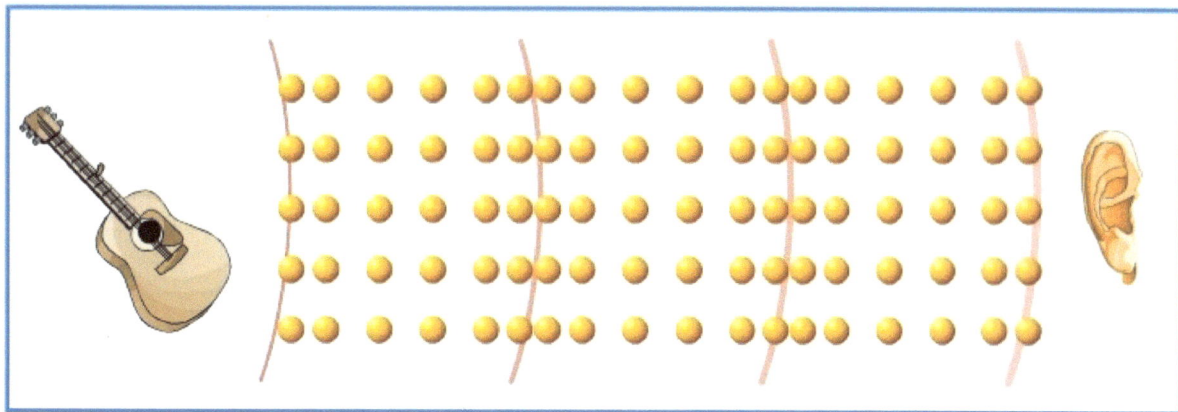

When an object vibrates, it causes vibrations in the air around it. These vibrations very quickly reach your ear and cause your eardrum (a soft thin soft part inside your ear) to vibrate. Signals are then sent to your brain and you hear sound.

     www.onboardacademics.com

Put a √ next to the objects that vibrate when they make a sound.

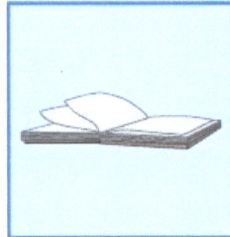

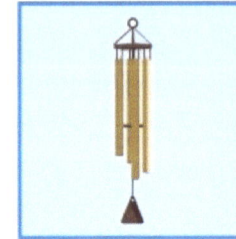

Can you arrange these objects by their sounds being softest to loudest? Draw the item in the correct box.

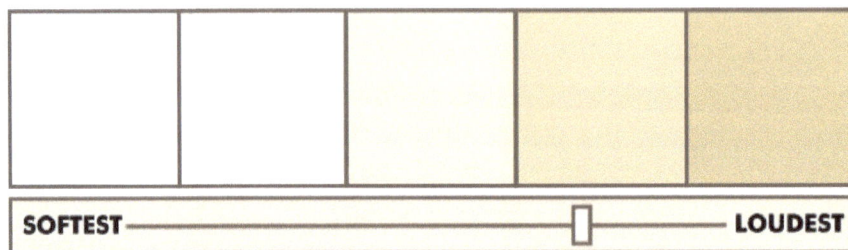

SOFTEST ———————————————————— LOUDEST

Objects that make loud sounds vibrate more than objects that make soft sounds. Loud sounds have more energy and so their vibrations travel further.

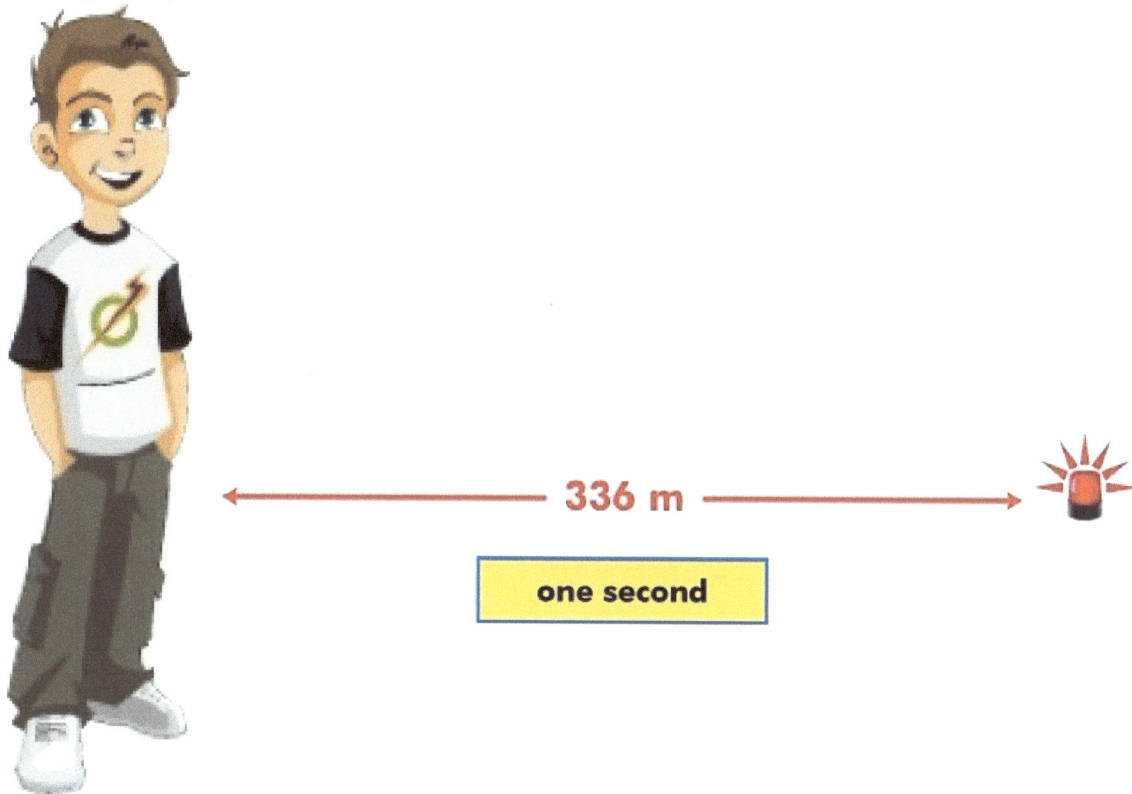

Owen is standing 336 meters(that's about one quarter of a mile) from a siren when it goes off. After the Owen sees the light from the siren, he won't hear the sound for one second.

The speed of sound changes depending on what material it's traveling through.

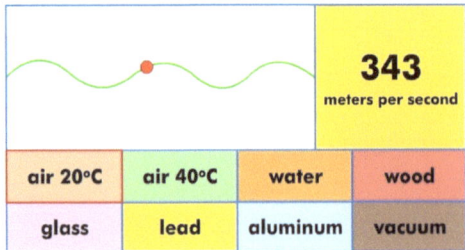

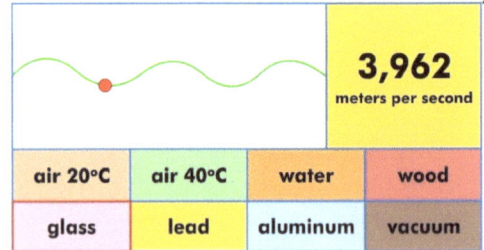

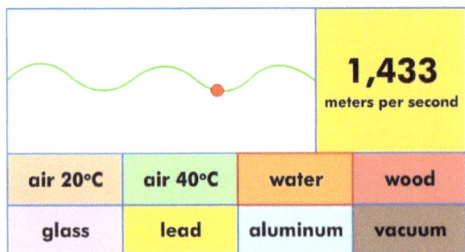

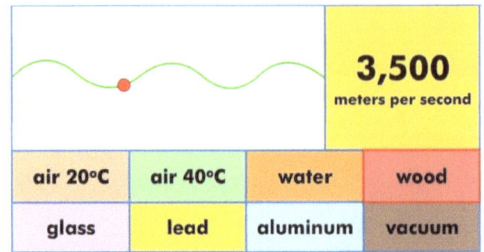

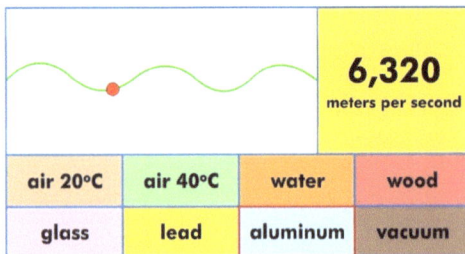

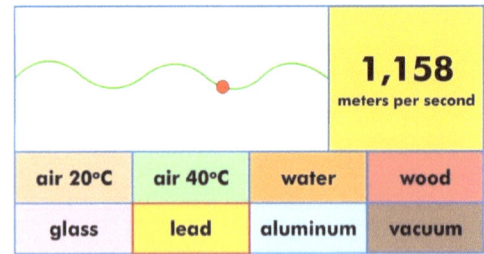

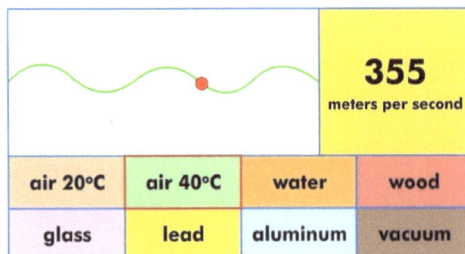

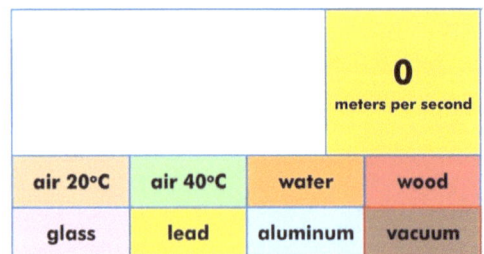

343 meters per second			
air 20°C	air 40°C	water	wood
glass	lead	aluminum	vacuum

3,962 meters per second			
air 20°C	air 40°C	water	wood
glass	lead	aluminum	vacuum

1,433 meters per second			
air 20°C	air 40°C	water	wood
glass	lead	aluminum	vacuum

3,500 meters per second			
air 20°C	air 40°C	water	wood
glass	lead	aluminum	vacuum

6,320 meters per second			
air 20°C	air 40°C	water	wood
glass	lead	aluminum	vacuum

1,158 meters per second			
air 20°C	air 40°C	water	wood
glass	lead	aluminum	vacuum

355 meters per second			
air 20°C	air 40°C	water	wood
glass	lead	aluminum	vacuum

0 meters per second			
air 20°C	air 40°C	water	wood
glass	lead	aluminum	vacuum

In which substance does sound travel the fastest? _____

In which substance does sound travel the slowest? _____

In which substance does sound not travel at all? _____

www.onboardacademics.com

Sound Quiz

1. Sound is a form of energy produced due to vibration. True or false?

2. Which of these objects make a sound; pencil, boiling water, book? _____

3. Vibration is when something moves back and forth quickly. True or false?

4. Which of the following is the softest sound?
   a. Phone ringing
   b. Butterfly fluttering
   c. Car honking

5. Apart from air, sounds can also travel through other materials. True or false?

6. Which does sound travel through faster, air or water?

   _____

7. Which is the loudest of the following sounds?
   a. Drums beating
   b. Leaves rustling
   c. Birds chirping

Properties of Light

www.onboardacademics.com

Do you know how long it takes light to travel from the sun to the earth?

Light travels amazingly fast. A beam of light could travel around the Earth over seven times in about a second. However, because the Sun is so far away, it actually takes sunlight about eight minutes to reach Earth.

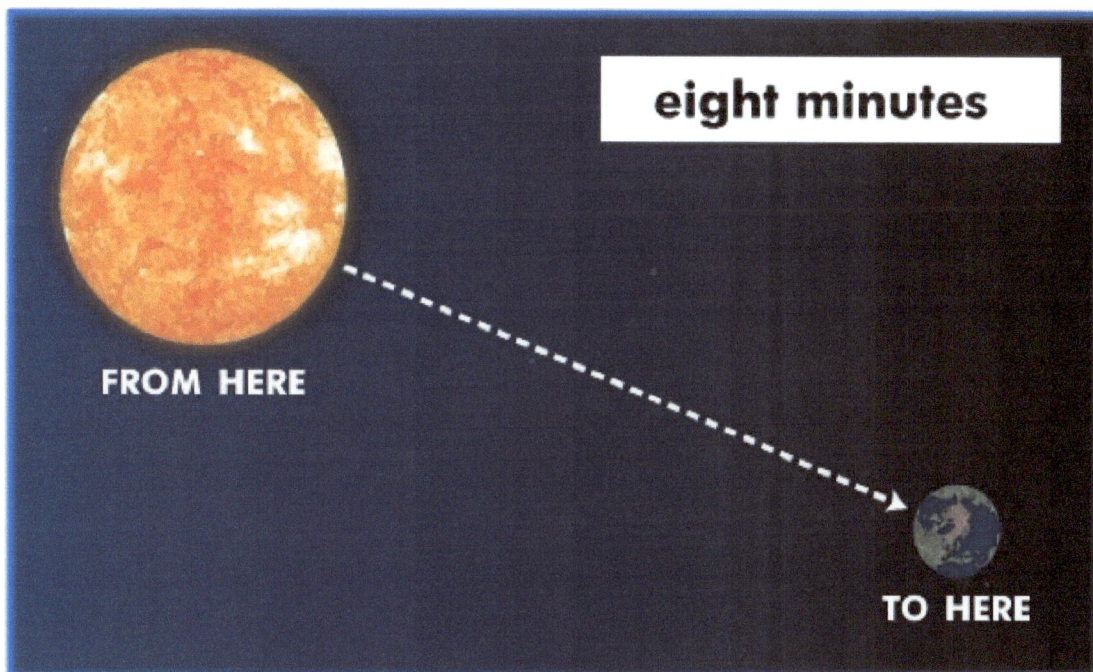

eight minutes

FROM HERE

TO HERE

The Source of Light

All light has a source and travels in a direction. If you can see light, then the light must have started somewhere and traveled to get there.

Circle the source of the light.

www.onboardacademics.com

Which direction will the light travel?

Draw light beams from each object in the direction that they will travel.

When traveling light is blocked it makes a shadow.

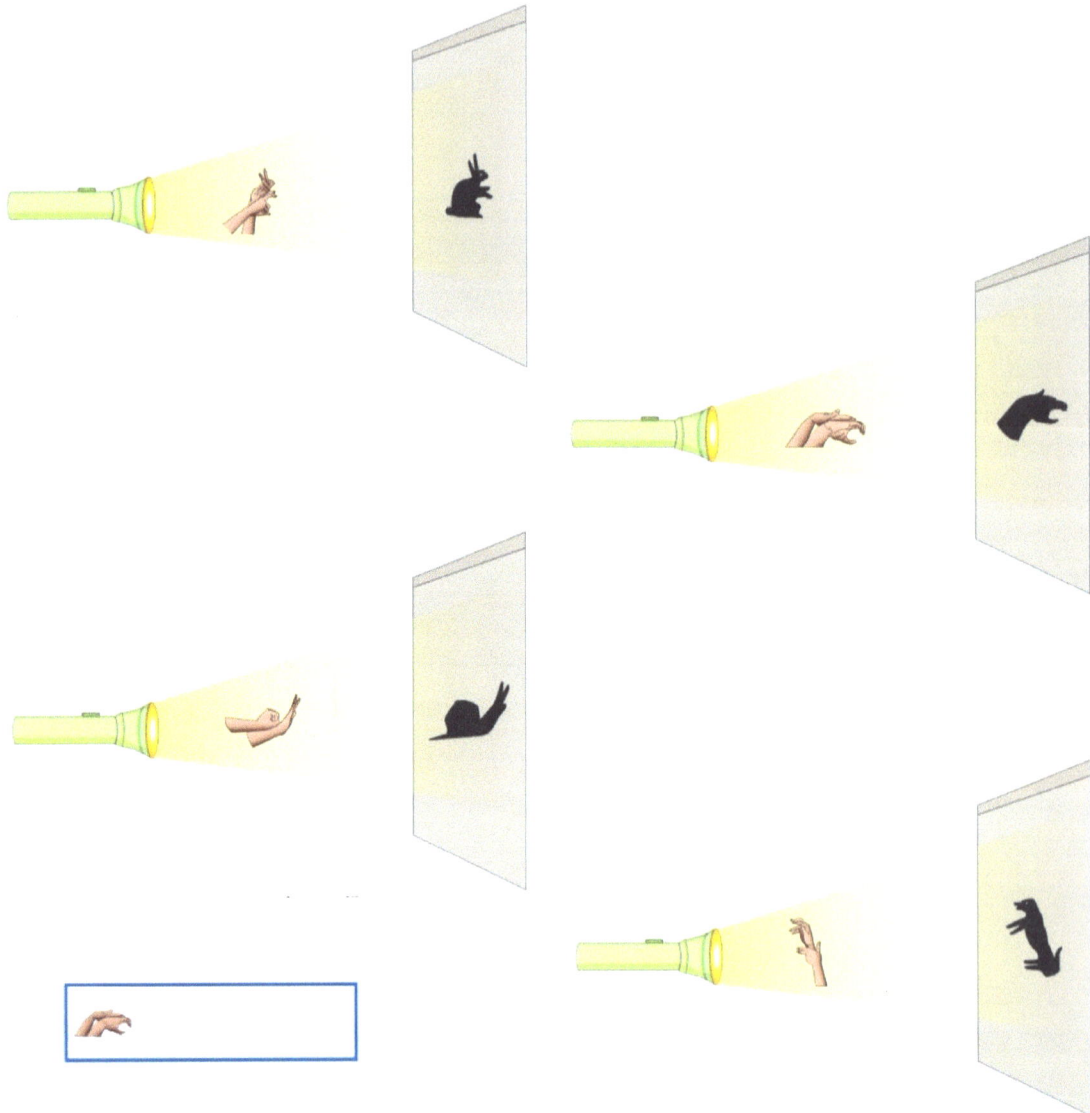

Can you name the shapes that these hand shadows make?

Light bounces or reflects off of objects.

When light travels from a light source and hits an object it bounces or reflects off of the object in many different directions. Some of the reflective light that bounces from the

object hits your eye entering through the pupil and then travels to a part of the eye called the retina. The retina sends a signal to the brain that processes the information and gives us the sensation of sight.

You only see the flower when light is reflected from the flower and hits your eye. If there is no light the flower is still there but you can't see it because there is no light to reflect off of the flower and into your eye.

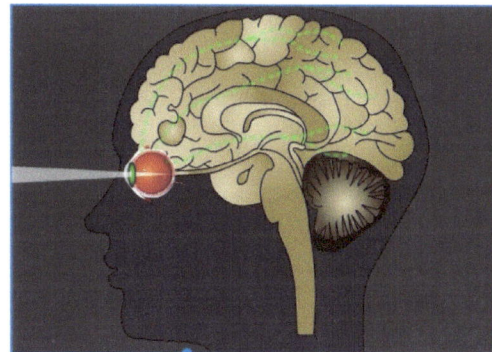

Reflection or bouncing light is why you can see yourself in a mirror. When you see yourself in the mirror, first the light reflects off of you and then reflects off of you and onto a mirror. Because mirrors are made with a very reflective material that makes almost all light reflect off of them, the light reflects off of the mirror and back into your eye. This is why you see your own image in the mirror.

So what it a reflection?

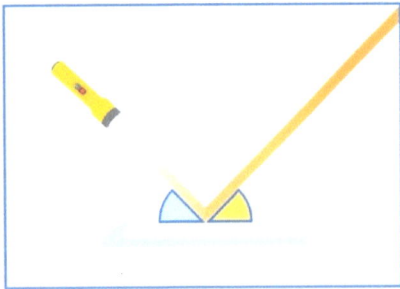

When light hits a shiny surface like a mirror, it gets reflected and bounces off in a new direction.

A mirror can be used to change the direction of a beam of light.

Sort these objects by how reflective they are.

Very reflective	Not very reflective

Properties of Light Quiz

Fill in the blanks.

All light has a _____ ; a place where the light comes from. Light doesn't just appear, it _____ in a direction. Light travels very _____. In fact, _____ is faster than light. If something blocks light, a _____ is created. When light hits an object, the light is _____ off the object. If the light reflects into your _____, then you can see the object. Some materials, like mirrors, are very _____.

absorbs	slowly	disappears	shadow
nothing	source	reflective	brain
quickly	travels	reflected	eyes

Light and Color

www.onboardacademics.com

How does a rainbow form?

You've probably seen a rainbow form when the sun comes out on a rainy day. But why does the combination of the sun and raindrops produce a rainbow?

What is your idea about why the sun and raindrops form a rainbow?

_____

_____

_____

Making a Rainbow with a Flashlight and a Prism

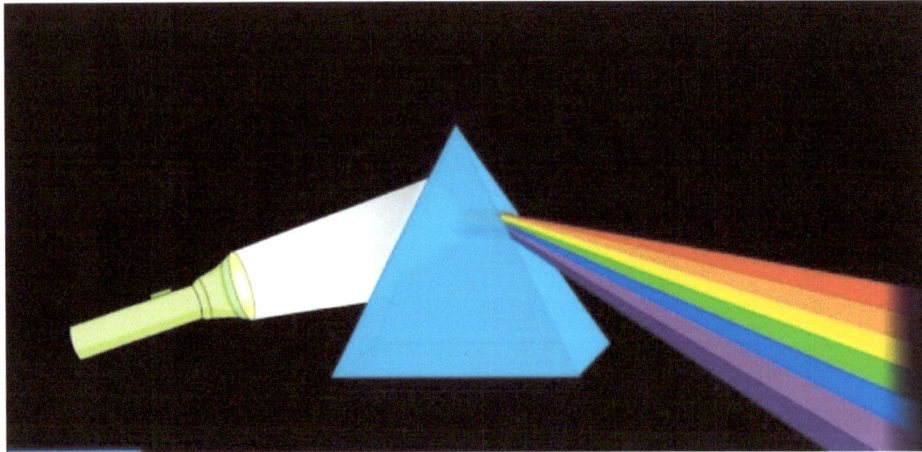

White light is actually made up of several different colors, and we can use a transparent prism to break up white light to observe the spectrum of colors within it. This is similar to observing a rainbow. Raindrops in the air act like tiny prisms that break up the light passing through them, creating the colorful rainbows that we see on the horizon.

Now that you've learned about prisms can you explain why sun and raindrops form a rainbow?

_____

_____

_____

www.onboardacademics.com

What is the visible light spectrum?

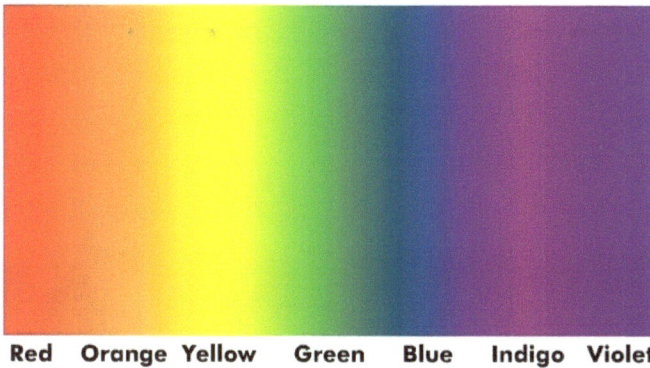

When broken apart, white light separates into the same spectrum of colors in order of wave length. The order of colors by wave length is; red, orange, yellow, green, blue, indigo and violet.

Red   Orange   Yellow   Green   Blue   Indigo   Violet

A good way to remember this is to think of an acronym such as Roy G. Biv.

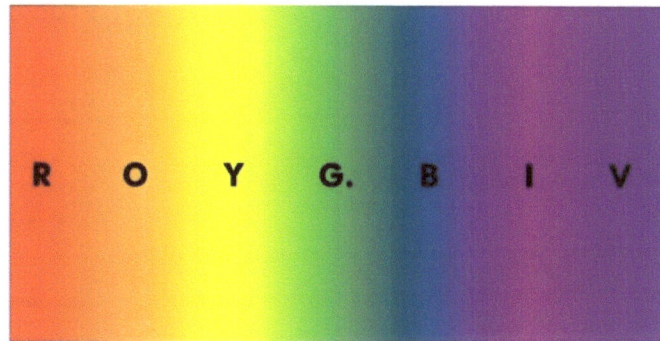

R   O   Y   G.   B   I   V

The wave length of a color tells you how much energy it has. For example red light has the least amount of energy and has the longest wave length. Violet has the most amount of energy and so it has the shortest wave length.

> **White light contains a spectrum of colors that appear in order of their wavelengths. From longest wavelength to shortest wavelength (or least energy to most energy), they are red, orange, yellow, green, blue, indigo and violet (or ROY G. BIV).**

www.onboardacademics.com

Order the spectrum of colors by wave length.

∿∿∿∿∿∿∿	Shortest	Indigo
∿∿∿∿∿∿		Orange
∿∿∿∿∿		Yellow
∿∿∿∿		Red
∿∿∿		Green
∿∿		Violet
∿	Longest	Blue

How is light related to color?

Objects appear to be a certain color because of the colors they absorb and the colors they reflect. For example look at this red rose.

When light is present, all colors of the spectrum are shining on the red rose; red, orange, yellow, green, blue, indigo and violet. The rose absorbs orange, yellow, green, blue, indigo and violet. However the red light is not absorbed and bounces off into our eye and that is why we see the rose as red.

When no light is present, there is no reflection of light to reflect off of objects into our eyes so nothing is visible.

Which colors are being observed and which colors are being reflected for each item.

X for absorbed

√ for reflected

	R	O	Y	G	B	I	V
🍃							
📘							
☂							
👕							

Primary Colors

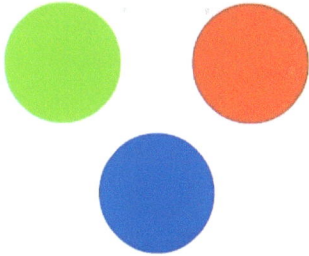

Almost any color can be created by mixing the three primary colors of light; red, green and blue.

For example when green and red overlap we see yellow. When all three colors overlap we see white light. This only works when mixing light.

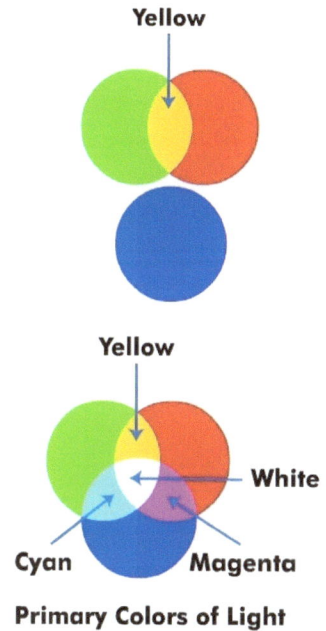

Yellow

Yellow

Cyan

White

Magenta

Primary Colors of Light

When mixing paint, the primary colors are yellow, cyan and magenta. When all three paint colors are mixed we get black.

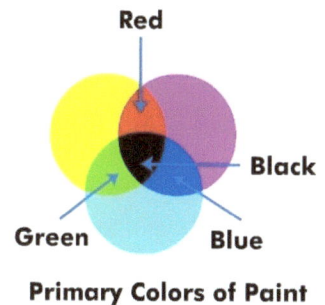

Red

Green

Black

Blue

Primary Colors of Paint

www.onboardacademics.com

What colors are created when you mix the following color combinations.

Light and Color Quiz

1. For a rainbow to be seen, you need light from the sun and raindrops in the air. True or false.

2. White light is made up of _____ colors
   a. 8
   b. 7
   c. 6
   d. 10

3. Raindrops in the air act like tiny prisms that break up light passing through them, creating rainbows. True or false?

4. When broken apart, white light splits into a spectrum of colors in order of _____.
   a. brightness
   b. wavelength

5. What is the acronym to remind you of the order of colors?

   _____

6. What color has the least energy? _____

7. What color has the most energy? _____

8. What color do you get when you mix the three primary colors of light? _____

www.ingramcontent.com/pod-product-compliance
Lightning Source LLC
Chambersburg PA
CBHW052049190326
41521CB00002BA/160